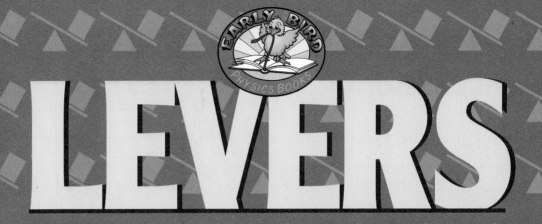

LEVERS

by Sally M. Walker and Roseann Feldmann

photographs by Andy King

Lerner Publications Company • Minneapolis

For my husband, Ron, love you forever—RF

*The publisher wishes to thank the Minneapolis Kids program
for its help in the preparation of this book.*

Additional photographs are reproduced with permission from: © *Leonard Lessin / Peter Arnold,
Inc., p. 16;* © *Caroline Penn / Corbis, p. 31.*

Lerner Publications Company
A division of Lerner Publishing Group
241 First Avenue North
Minneapolis, MN 55401 U.S.A.

Website address: www.lernerbooks.com

Library of Congress Cataloging-in-Publication Data

Walker, Sally M.
  Levers / by Sally M. Walker and Roseann Feldmann ;
photographs by Andy King.
     p.   cm. — (Early bird physics books)
Includes index.
  ISBN 0-8225-2218-7 (lib. bdg. : alk. paper)
  1. Levers—Juvenile literature. [1. Levers.] I. Feldmann,
Roseann. II. King, Andy, ill. III. Title. IV. Series
TJ147.W36 2002
621.8'11—dc21   00-010740

Manufactured in the United States of America
2 3 4 5 6 7 – JR – 07 06 05 04 03 02

CONTENTS

BE A WORD DETECTIVE

Can you find these words as you read about levers?
Be a detective and try to figure out what they mean.
You can turn to the glossary on page 46 for help.

complicated machines   **load**
first-class lever   **second-class lever**
force   **simple machines**
fulcrum   **third-class lever**
lever   **work**

You do work when you write. What does the word "work" mean to a scientist?

Chapter 1

WORK

You work every day. You do chores around your home. At school you write. It may surprise you to learn that playing and eating are work, too!

When scientists use the word "work," they don't mean the opposite of play. Work is using force to move an object from one place to another. Force is a push or a pull. You use force to carry out the trash. And you use force to turn the page of a book.

This girl is using a bat to move a rock. She is using force to move it, so she is doing work.

Every time you use force, the force has a direction. Force can move in any direction. When you push open a door, the force is aimed away from you.

▲ *The direction of a force can be away from you.*

You use a downward force when you type.

You use an upward force to open some kinds of windows. And you use a downward force when you type on a computer.

Every time your force moves an object, you have done work. It doesn't matter how far the object moves. If it moves, work has been done. Throwing a ball is work. Your force moves the ball from one place to another.

You do work when you move sand from one place to another place.

These kids are pushing hard. But they have done no work.

Pushing your school building is not work. It's not work if you sweat. It's not work even if you push until your arms feel like rubber. No matter how hard you push, you haven't done work. The building hasn't moved. If the building moves, then you worked!

A vacuum cleaner is a machine that has many moving parts. What kind of machine is it?

Chapter 2
MACHINES

Most people want their work to be easy. Machines are tools that make work easier.

Some machines have many moving parts. These machines are called complicated machines. Cars and vacuum cleaners are complicated machines.

A light switch is a simple machine.

Some machines have only a few moving parts. These machines are called simple machines. Simple machines are found in every home, school, and playground. They are so simple that you might not realize they are machines.

Simple machines make work easier in many ways. One way is by changing the direction of force. When you use your arms to lift a friend, you use an upward force. But you can lift your friend more easily by using a downward force. How? If she sits on a seesaw, the end she sits on goes down. When you sit on the other end, your friend goes up. Your force is downward. But your friend still goes up.

This girl sits on one end of the seesaw. The other end of the seesaw goes up and lifts her friend.

A bottle opener is a simple machine called a lever. What do levers help people to do?

Chapter 3

PARTS OF A LEVER

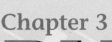

You can use a seesaw to lift a friend. The seesaw is a simple machine. This kind of simple machine is called a lever. A lever is a bar that is hard to bend. Levers make it easier to move things.

A lever must rest on another object. The object a lever rests on is called its fulcrum.

You can make a lever. You will need a 12-inch wooden ruler, a crayon, a small can of food, and some rubber bands.

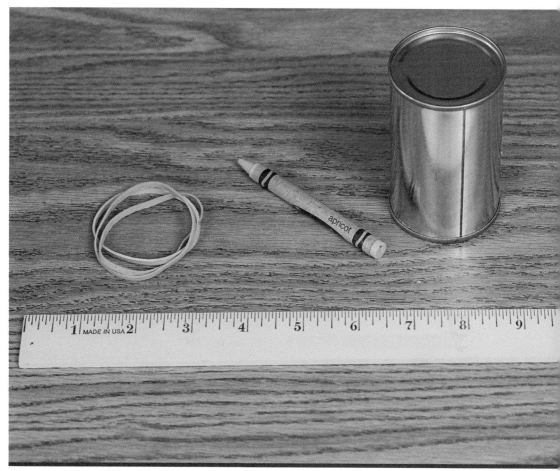

You can use these objects to make your own lever.

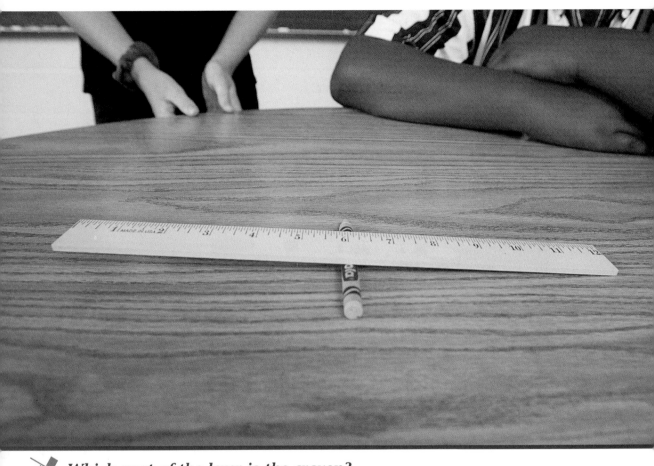

Which part of the lever is the crayon?

Place the ruler on the crayon. Put the crayon under the ruler's 6-inch mark. One end of the ruler will probably touch the table. The ruler is your lever. It rests on top of the crayon. So the crayon is the ruler's fulcrum.

18

The ruler and the crayon work together. Push down on the high end of the ruler. What happens? Your downward force makes the other end of the ruler go up.

When you push down on one end of the lever, the other end goes up.

Put one finger on each end of the ruler.
Push one end down. Then push the other end
down. Watch the crayon. What happens? The
crayon stays in the same place while the ruler
moves around it.

*The crayon is the lever's fulcrum. The lever
moves, but the fulcrum stays in the same place.*

Now put the can on the ruler. The middle of the can should be on top of the ruler's 11-inch mark. Use the rubber bands to tie the can to the ruler. The can is the lever's load. A load is an object you want to move.

The can is the lever's load.

A lever helps you lift a load.

Put the crayon under the ruler's 6-inch mark. Push the high end of the ruler down. Your force makes the lever move around its fulcrum. Lifting the load is easy. You don't have to use much force.

Think about your lever. Your finger makes a force on one end of the lever. The can is the load at the other end of the lever. The crayon is the fulcrum between the load and the force.

A lever can't hold up a load without a force. If you stop pushing down on the ruler, the can crashes down.

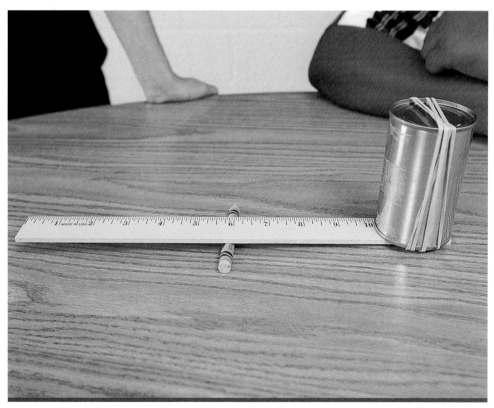

A lever can't hold up a load without force.

The crayon is in a new place now. Will moving the crayon change how hard you have to work?

Chapter 4

CHANGING THE AMOUNT OF FORCE

You can change how much force you must use to lift the can. To change the force you must change the lever a bit.

Put the crayon under your lever's 9-inch mark. The lever looks different now. The fulcrum is far away from your force. Push down on the high end. It's easy to lift the load. You need to use only a little force. Moving the fulcrum farther away from the force makes your work easier.

Next, put the crayon under the ruler's 3-inch mark. Now the fulcrum is close to your force. Push down on the high end of the lever. You have to use a lot of force to lift the load. Putting the fulcrum close to the force makes your work harder.

When the fulcrum is close to the force, your work is harder.

Does moving the fulcrum change how high the load is lifted? Put the crayon under the ruler's 9-inch mark. Notice how high the ruler's end is above the tabletop.

The fulcrum is close to the load again.

The load is raised only a small distance.

Push down on the lever to lift the load. When the can goes up, try to slide a finger under the raised end of the ruler. There's probably just enough room for your finger to fit. Your long downward push is easy. But it raises the can only a small distance.

This time, the fulcrum is far from the load.

Put the crayon back under the ruler's 3-inch mark. The end you will push down is much closer to the tabletop now. What happens to the load when you push down again?

The load goes up high when the fulcrum is far from the load. But lifting the load is hard.

The can goes up higher this time. You may be able to fit two fingers under the ruler's end. Your short downward push was a lot harder. But it lifted the can much higher.

Press the ruler's end halfway down. Both ends of the ruler should be above the table. Now push the ruler back and forth over the crayon. Notice how your force changes as you do this. Your force changes as your finger gets closer to or farther from the fulcrum.

As your finger moves closer to the fulcrum, you need to use more force. As your finger moves farther from the fulcrum, you need less force.

This girl is using a water pump. The handle is a lever.

When you use a lever, ask yourself two questions. Do you want to use a little force and move the load a little bit? Or do you want to use a lot of force and move the load a lot? Your answer will help you decide where to put the fulcrum. If the fulcrum is far from the force, the load moves a little. So you only need a little force. If the fulcrum is close to the force, the load moves a lot. But you must use a lot of force.

This girl is trying to lift the lid of a paint can. She is using a screwdriver as a lever. How many kinds of levers are there?

Chapter 5
KINDS OF LEVERS

There are three kinds of levers. The lever you made from a ruler is one kind of lever. It is called a first-class lever. In a first-class lever, the fulcrum is between the load and the force. A hammer is a first-class lever when you use it to pull out a nail. The nail is the load. The person pulling makes the force. And the fulcrum is the place where the hammer's head rests on the board. The fulcrum is between the load and the force.

You can use a hammer to pull out a nail. When you do this, the hammer is a first-class lever.

You can make your ruler into a second-class lever. The end of the ruler should hang over the edge of the table.

The second kind of lever is called a second-class lever. In a second-class lever, the load is between the fulcrum and the force.

You can make your ruler into a second-class lever. Make sure the can is still attached at the ruler's 11-inch mark. Lay the ruler on the table. Put the 1-inch mark at the edge of the table. Most of the ruler will be on the table. But 1 inch will hang over the edge.

Lift the end of the ruler several inches. Look at the lever. Can you find the fulcrum? The lever is resting on the tabletop. So the tabletop is the fulcrum. The load is between the fulcrum and the force.

In a second-class lever, the load is between the fulcrum and the force.

Move the middle of the can to the ruler's 6-inch mark. Lift the end the same amount as you did before. Then move the can to the 3-inch mark and try it again. Do you need to use more force when the load is closer to your hand?

It is harder to lift a load when it is close to the force.

Notice how high the can is lifted each time. When the load is far from the force, the load moves only a little. But it's easy to lift. When the load is close to the force, the load moves a lot. But you need a lot of force to lift it.

The load moves a lot when it is close to the force.

A wheelbarrow is a second-class lever. The wheel is the fulcrum. The force is at the handles. The load is inside the wheelbarrow. The load is between the fulcrum and the force. When the load is toward the front of the wheelbarrow, it is easy to lift. When the load is closer to the handles, it is harder to lift.

A wheelbarrow is a second-class lever.

The third kind of lever is called a third-class lever. A third-class lever has the force between the fulcrum and the load.

A broom is a third-class lever. You hold the broom in two places. Your bottom arm gives the force. The top arm on the broom is the fulcrum. The dirt is the load.

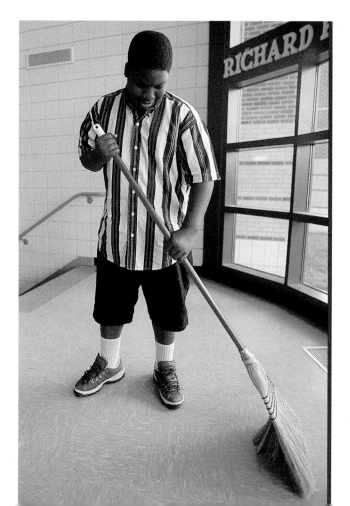

A broom is a third-class lever.

When you kick a ball, your leg is a third-class lever. Can you tell where the fulcrum, load, and force are?

A third-class lever helps you move objects a long distance. A good sweep makes the broom move a long distance. You can reach out and move a lot of dirt easily.

KINDS OF LEVERS

FIRST–CLASS LEVER: the fulcrum is between the load and the force

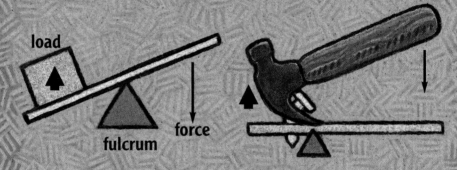

SECOND–CLASS LEVER: the load is between the fulcrum and the force

THIRD–CLASS LEVER: the force is between the load and the fulcrum

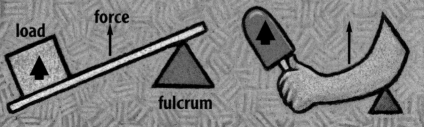

Levers make doing work easier. Some levers increase your force. Some levers change the direction of your force. And some levers help you move an object a long distance.

Pruning shears are two levers held together with a bolt. The load is far from the force. So using pruning shears makes a cutter's work easier.

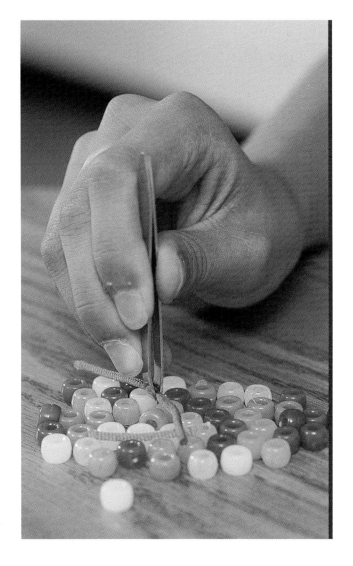

Tweezers help you move small loads easily. A pair of tweezers is two levers put together. The force is between the fulcrum and the load. What kind of lever is a pair of tweezers?

Using a lever gives you an advantage. An advantage is a better chance of finishing your work. Using a lever is like having a helper. The work is easier. And that's a real advantage!

ON SHARING A BOOK

When you share a book with a child, you show that reading is important. To get the most out of the experience, read in a comfortable, quiet place. Turn off the television and limit other distractions, such as telephone calls. Be prepared to start slowly. Take turns reading parts of this book. Stop occasionally and discuss what you're reading. Talk about the photographs. If the child begins to lose interest, stop reading. When you pick up the book again, revisit the parts you have already read.

Be a Vocabulary Detective
The word list on page 5 contains words that are important in understanding the topic of this book. Be word detectives and search for the words as you read the book together. Talk about what the words mean and how they are used in the sentence. Do any of these words have more than one meaning? You will find the words defined in a glossary on page 46.

What about Questions?
Use questions to make sure the child understands the information in this book. Here are some suggestions:

> What did this paragraph tell us? What does this picture show? What do you think we'll learn about next? What is force? Can force move in any direction? How are simple machines different from complicated machines? How do levers help people? What is the object a lever rests on called? How many kinds of levers are there? What is your favorite part of the book? Why?

If the child has questions, don't hesitate to respond with questions of your own, such as: What do *you* think? Why? What is it that you don't know? If the child can't remember certain facts, turn to the index.

Introducing the Index
The index helps readers find information without searching through the whole book. Turn to the index on page 47. Choose an entry such as *load* and ask the child to use the index to find out what a lever's load is. Repeat with as many entries as you like. Ask the child to point out the differences between an index and a glossary. (The index helps readers find information, while the glossary tells readers what words mean.)

LEARN MORE ABOUT
SIMPLE MACHINES

Books

Baker, Wendy, and Andrew Haslam. *Machines*. New York: Two-Can Publishing Ltd., 1993. This book offers many fun educational activities that explore simple machines.

Burnie, David. *Machines: How They Work*. New York: Dorling Kindersley, 1994. Beginning with descriptions of simple machines, Burnie goes on to explore complicated machines and how they work.

Hodge, Deborah. *Simple Machines*. Toronto: Kids Can Press Ltd., 1998. This collection of experiments shows readers how to build their own simple machines using household items.

Van Cleave, Janice. *Janice Van Cleave's Machines: Mind-boggling Experiments You Can Turn into Science Fair Projects*. New York: John Wiley & Sons, Inc., 1993. Van Cleave encourages readers to use experiments to explore how simple machines make doing work easier.

Ward, Alan. *Machines at Work*. New York: Franklin Watts, 1993. This book describes simple machines and introduces the concept of complicated machines. Many helpful experiments are included.

Websites

Brainpop—Simple Machines
<http://www.brainpop.com/tech/simplemachines/> This site has visually appealing pages for levers and inclined planes. Each page features a movie, cartoons, a quiz, history, and activities.

Simple Machines
<http://sln.fi.edu/qa97/spotlight3/spotlight3.html> With brief information about all six simple machines, this site provides helpful links related to each and features experiments for some of them.

Simple Machines—Basic Quiz
<http://www.quia.com/tq/101964.html> This challenging interactive quiz allows budding physicists to test their knowledge of work and simple machines.

GLOSSARY

complicated machines: machines that have many moving parts

first-class lever: a lever that has its fulcrum between the load and the force

force: a push or a pull

fulcrum: the object a lever rests on

lever: a stiff bar that is used to move other objects

load: an object you want to move

second-class lever: a lever that has its load between the fulcrum and the force

simple machines: machines that have few moving parts

third-class lever: a lever that has its force between the fulcrum and the load

work: moving an object from one place to another

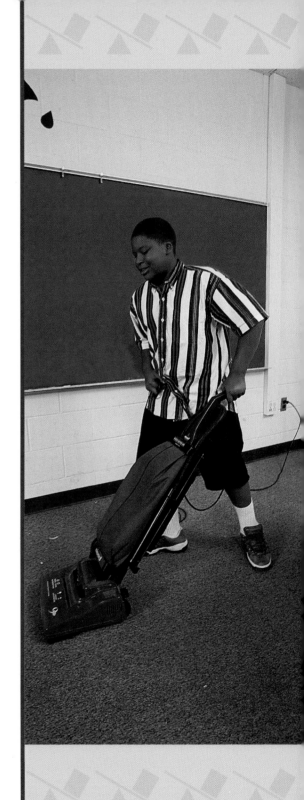

INDEX

Pages listed in **bold** type refer to photographs.

About the Authors

Sally M. Walker is the author of many books for young readers. When she isn't busy writing and doing research for books, Ms. Walker works as a children's literature consultant. She has taught children's literature at Northern Illinois University and has given presentations at many reading conferences. She lives in Illinois with her husband and two children.

Roseann Feldmann earned her B.A. degree in biology, chemistry, and education at the College of St. Francis and her M.S. in education from Northern Illinois University. As an educator, she has been a classroom teacher, college instructor, curriculum author, and administrator. She currently lives on six tree-filled acres in Illinois with her husband and two children.

About the Photographer

Freelance photographer Andy King lives in St. Paul, Minnesota, with his wife and daughter. Andy has done editorial photography, including several works for Lerner Publishing Group. Andy has also done commercial photography. In his free time, he plays basketball, rides his mountain bike, and takes pictures of his daughter.

METRIC CONVERSIONS

WHEN YOU KNOW:	MULTIPLY BY:	TO FIND:
miles	1.609	kilometers
feet	0.3048	meters
inches	2.54	centimeters
gallons	3.787	liters
tons	0.907	metric tons
pounds	0.454	kilograms